S0-AQG-916

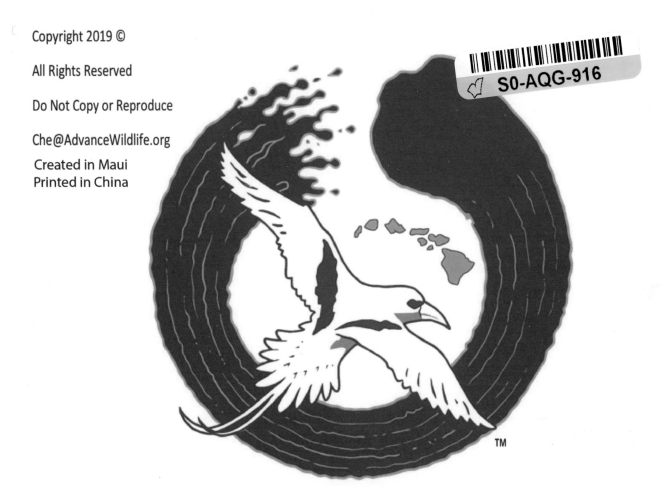

TM

Advance Wildlife Education

AdvanceWildlifeEducation.org

AWE was founded in order to build a bridge between the public and conservation organizations and nonprofits, through education, art, jewelry, and attire.

Wildlife Biologist, Founder, and Illustrator

Che Frausto

Dedicated to my parents and family

Tyrannosaurus Rex

Fun Facts

- The name Tyrannosaurus Rex comes from Greek and Latin words meaning Tyrant Lizard King.
- The T. Rex roamed the planet during the late Cretaceous Period 65 million years ago, in what's now the USA and Canada.
- The T. Rex grew up to 40 feet long and 20 feet tall.
- These animals had a set of 50-60 banana-sized teeth they could rip almost 220 lbs. of flesh off in a single bite.
- T. rex's teeth were wide and somewhat dull (rather than being flat and daggerlike), allowing the teeth to withstand the forces exerted by struggling prey.
- They had the strongest bite of any land animal that ever lived, according to a 2012 study in the journal Biology Letters. The dinosaur's bite could exert up to 12, 814 pounds-force (57,000 Newtons), which is roughly equivalent to the force of a medium-size elephant sitting down.
- The T. Rex was super smart too, boasting a brain twice as big as those of the other giant carnivores. The T. rex was roughly as intelligent as a chimp.
- Scientist believe the Tyrannosaurus Rex could run at speeds of up to 25 mph.
- This dinosaur's tiny two fingered clawed arms (less than one meter in length) remain a bit of a mystery to this day.
- More than 20 almost-complete T rex skeletons have been found. The best skeleton nicknamed Sue was unearthed in South Dakota, USA.
- In 1997, this super skeleton was sold to a museum in Chicago for around 5 million dollars.

Diet

- T. rex was a huge carnivore and primarily ate herbivorous plant eating dinosaurs. They were probably opportunistic and may have fed on carcasses as well as hunt live prey.
- Scientists are unsure whether T. rex hunted alone or in packs. In 2014, researchers found dinosaur track marks in the foothills of the Canadian Rockies in British Columbia — out of the seven tracks, three belonged to Tyrannosaurids.

Triceratops

Fun Facts

- Triceratops, with its three horns and bony frill around the back of its head, is one of the most recognizable dinosaurs.
- Its name is a combination of the Greek syllable's tri-, meaning "three," kéras, meaning "horn," and ops, meaning "face."
- The dinosaur roamed North America about 67 million to 65 million years ago, during the end of the Cretaceous Period.
- Triceratops were first discovered in 1887.
- Triceratops were massive animals, comparable in size to African elephants. According to a 2011 article in the journal Cretaceous Research. It grew up to 30 feet, 10 feet tall and weighed well over 11,000 lbs.— some large specimens weighed nearly 15,750 lbs.
- The head of Triceratops was among the largest of all land animals, some making up one-third of the entire length of the dinosaur's body. The largest skull found has an estimated length of 8.2 feet.
- During a Triceratops' juvenile years, its horns were little stubs that curved backward; as the animal continued to grow into young adulthood, the horns straightened out; finally, the horns curved forward and grew up to 3 feet long (1 meter), probably after the dinosaur reached sexual maturity.
- The dinosaur also used its horns and frill in fights against its main predator, tyrannosaurs. Paleontologists have uncovered brow horn and skull Triceratops bones that were partially healed from tyrannosaur tooth marks, suggesting the Triceratops successfully fended off its attacker.

Diet

- The Triceratops was an herbivore, existing mostly on shrubs and other plant life. Its beak-like mouth was best suited for grasping and plucking rather than biting, according to a 1996 analysis in the journal Evolution. It also likely used its horns and bulk to tip over taller plants.
- It had up to 800 teeth that were constantly being replenished and were arranged in groups called batteries.

Utahraptor

Fun Facts

- Utahraptor was by far the biggest raptor ever to walk the earth.
- This group of carnivorous dinosaurs had a large retractable sickle claw on its foot, specialized for cutting.
- With a name meaning "Utah's predator," Utahraptor was a ferocious hunter that used its sickle-shaped claws to attack and rip apart its prey.
- The claw itself was about 9.5 inches long.
- Utahraptor was the inspiration for the Velociraptors in the film Jurassic Park.
- It was bigger than the Velociraptor; adults were around 20 feet long and around 5 feet tall at the hip.
- They weighed up 2,200 lbs.
- Utahraptor is closely related to birds.
- Raptors of the late Cretaceous period, like Deinonychus and Velociraptor, were covered with feathers, at least during certain stages of their life cycles.
- Utahraptor was almost certainly warm-blooded, which would have been bad news for its presumably cold-blooded, plant-eating prey.

Diet

- They were carnivorous and primarily ate herbivorous dinosaurs.

Argentinosaurus

Fun Facts

- The largest land animal ever found, Argentinosaurus evolved in the Cretaceous Period.
- It was discovered in Argentina in 1987.
- Some reconstructions put this dinosaur at 75 to 85 feet from head to tail and up to 220,000 lbs. (about 17 African elephants).
- Given its enormous size, this dinosaur walked along at a top speed of five miles per hour, probably inflicting plenty of collateral damage along the way.
- Argentinosaurus would have laid 10-15 eggs at a time about the size of a coconut. Its hatchlings were born weighing just 11 lbs. and each one would have taken over 40 years to reach adult size.
- Argentinosaurus would probably have had weak jaws with pencil-like teeth for stripping the leaves off conifer trees. Unfortunately, we do not know for sure because no skull has ever been found.
- Argentinosaurus probably had scaly armor, like modern-day crocodiles and armadillos.

Diet

- Argentinosaurus would have eaten plants including prehistoric conifers. It is thought that this massive dinosaur would have had to ingest 100,000 calories a day to reach its adult size – that's the equivalent of 50 whole chocolate cakes or 2,127 apples.

Kronosaurus

Fun Facts

- The genus name Kronosaurus, meaning 'Kronos lizard', was named after the Greek titan Kronos.
- Kronosauruses belongs to an extinct group of marine reptiles known as the Plesiosauria, they generally had short necks, broad bodies, four flippers, and short tails.
- The last known plesiosaurs perished during a mass extinct event at the end of the Late Cretaceous 65 million years ago.
- Kronosaurus was a relatively large plesiosaur, measuring approximately 36 feet in length and weighing 24,250 lbs.
- The largest recorded Kronosaurus skull measures 7.2 feet in length.
- Estimated bite force for the skull of Kronosaurus is estimated up to 30,000 newtons — approximately twice as powerful as a large saltwater crocodile.
- The teeth of Kronosaurus are uneven in shape and size throughout the jaws. Enormous fang-like teeth can measure up to a foot long from the tip to the bottom of the root.
- Four strong, paddle-like limbs would have propelled Kronosaurus through the water. The front limbs could span approximately 7 feet in diameter.
- Based on evidence from other plesiosaurs, Kronosaurus would have given birth to live young underwater. Baby Kronosaurus would have been birthed headfirst to prevent drowning before swimming to the surface to breath. Like modern whales, Kronosaurus was adapted for life at sea and would be immobilized underneath its own weight if stranded on land.

Diet

- Kronosaurus queenslandicus was a predator with a varied diet. The fossilized stomach contents of some specimens contain the bones of other marine reptiles such as small sea turtles. Shark vertebrae have also been preserved with a Kronosaurus specimen, possibly indicating a predator-prey relationship. It is likely that Kronosaurus also ate smaller prey items such as fish and cephalopods.

Stegosaurus

Fun Facts

- Stegosaurus means "roofed lizard," which was derived from the belief by 19th-century paleontologists that the plates lay flat along its back like shingles on a roof.
- Stegosaurus was a large, plant-eating dinosaur that lived during the late Jurassic period, about 150.8 million to 155.7 million years ago, primarily in western North America.
- Stegosaurus would have defended itself from predators with its powerful spiked tail.
- The largest species was Stegosaurus armatus, it grew up to about 30 feet long.
- The bony plates along its back were embedded in the skin of the animal, not attached to its skeleton, which is why in most fossil finds the plates are separated from the body.
- When the first fossil of a Stegosaurus was described, they concluded that the plates would have lain flat on its back. After finding a specimen that had been covered with mud, which had held the plates in place, Marsh realized that they stood vertically, alternately on either side of the spine.
- Scientists are not exactly sure what the plates were used for. They may have warned off predators or allowed members of the same species to recognize each other. Another suggestion is that the plates were used to regulate body temperature.
- Stegosaurus has a reputation for having a small brain and one of the lowest-brain-to-body ratios among dinosaurs. "The brain of Stegosaurus was long thought to be the size of a walnut," said armored dinosaur expert Kenneth Carpenter, director of the USU Eastern Prehistoric Museum in Utah. "But actually, its brain had the size and shape of a bent hotdog."

Diet

- Stegosaurus was an herbivore, as its toothless beak and small teeth were not designed to eat flesh and its jaw was not very flexible.
- Interestingly, unlike other herbivorous, beaked dinosaurs (including Triceratops and the duck billed Hadrosaurids), Stegosaurus did not have strong jaws and grinding teeth.
- Instead, its jaws likely only allowed up and down movements, and its teeth were rounded and peg like. It also had cheeks, which gave it room to chew and store more food than many other of dinosaurs.

Pteranodon

Fun Facts

- Flying reptile (pterosaur) fossils found in North American dated back 90 to 100 million years ago during the Late Cretaceous Period.
- Pteranodon had a wingspan of 23 feet or more, and its toothless jaws were very long and pelican-like.
- A crest at the back of the skull (a common feature among pterosaurs) may have functioned in species recognition; the crest of males was larger. The crest is often thought to have counterbalanced the jaws or have been necessary for steering in flight.
- Although the limbs appear tough, the bones were completely hollow, and their walls were no thicker than about .4 inch.
- The eyes were relatively large, and the animal may have relied heavily upon sight as it searched for food above the sea.
- Paleontologists speculate that it may have skimmed the water while in flight, capture fish near the water's surface, or dove after prey as modern diving seabirds do.
- Fossils of Pteranodon and related forms are found in Europe, South America, and Asia in rocks formed from substances found in marine environments, which supports the inference of a pelican-like lifestyle.

Diet

- The design of Pteranodon's jaws and the discovery of fossilized fish bones and scales with Pteranodon specimens suggest that it was a fish eater.

Spinosaurus

Fun Facts

- Spinosaurus was the biggest of all the carnivorous dinosaurs, larger than the Tyrannosaurus and Giganotosaurus.
- It lived during part of the Cretaceous period, about 112 million to 97 million years ago, roaming the swamps of North Africa.
- Spinosaurus means "spine lizard," an appropriate descriptor, as the dinosaur had very long spines growing on its back to form what is referred to as a "sail".
- The distinctive spines, which grew out of the animal's back vertebrae, were up to 7 feet long and were likely connected to one another by skin.
- Recent fossil evidence shows Spinosaurus was the first dinosaur that was able to swim, and likely spent most of its life in the water.
- Spinosaurus had short hind limbs (like early whales and other animals that spent more and more time in the water, wide and flat claws and feet (possibly used in paddling), and a long and slender snout with conical teeth (perfect for catching fish).
- Researchers in multiple studies in the Journal of Vertebrate Paleontology estimated Spinosaurus was 41 to 59 feet long and weighed 15,400 to 46,000 lbs.

Diet

- While its jaw was powerful, none of the teeth were serrated, making it unlikely that it could have used them to tear into tough prey. This gives credence to the theory that it mostly survived on fish and carcasses.
- Spinosaurus is thought to have survived primarily on fish, including giant coelacanths, sawfish, large lungfish, and sharks.

Magnapaulia

Fun Facts

- It lived in the Cretaceous period and inhabited North America. Its fossils have been found in places such as Baja California (Mexico).
- They existed from 83.5 million years ago to 70.6 million years ago.
- Magnapaulia laticaudus was estimated at 41 feet long and weighing 12,000 lbs.
- Maganpaulia may have spent a lot of time in the water.

Diet

- Plants

Dunkleosteus

Fun Facts

- 358 million years ago this was the largest predator and one of the fiercest creatures alive in the Devonian period "Age of Fishes,".
- Up to 32 feet in length and weighing more than 1 ton (2000 lbs), this fish could chop prehistoric sharks into chum.
- Dunkleosteus had a massive skull made of thick, bony plates, and 2 sets of fang-like protrusions near its powerful jawbones.
- Dunkleosteus did not have true teeth; instead, the skull's bony plates extended into sharpened "fangs" in front of the mouth. These fangs scraped together, continuously sharpening each other as the fish opened and closed its jaws.
- They found that as these monstrous fish grew up, their mouths changed. The jaws gradually lengthened, while the fangs up front grew sturdier. This meant the jaws of an adult closed more slowly, but with a lot more power. While the younger fish were eating smaller, softer prey, the adults were capable of punching through even other heavily armored arthrodires.

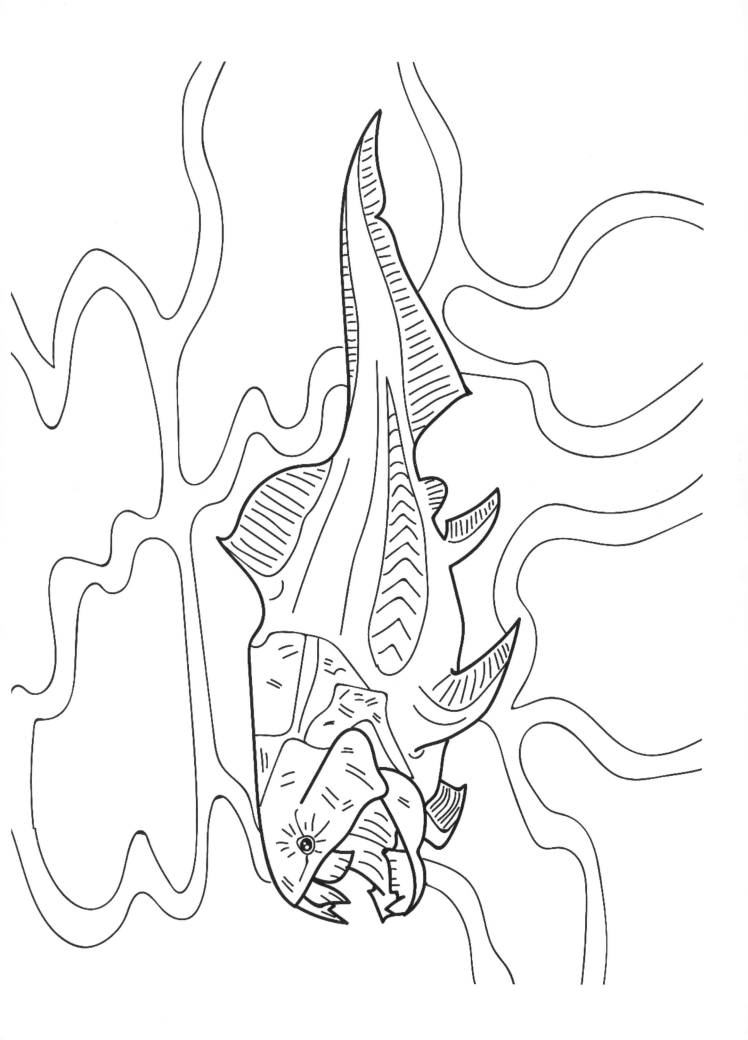

Apatosaurus

Fun Facts

- Apatosaurus was an herbivorous sauropod dinosaur that lived from about 155.7 to 150.8 million years ago.
- It's believed to be one of the biggest land animals to have roamed the Earth.
- Fossils suggest the dinosaur reached 69 to 75 feet in length. In the past, Apatosaurus was estimated to weigh up to 39 tons (78,000 lbs), but modern modeling techniques puts its average weight closer to 19.8 tons (39,600 lbs), according to a 2009 study in the Journal of Zoology.
- Like other sauropods, the vertebrae were made up of paired spines, producing a very wide and thick neck. But its neck was not as heavy as it might have been, thanks to a system of air sacs that kept it relatively light for its size.
- Apatosaurus had only a single large claw on each forelimb, with the first three toes on the hind limb possessing claws.
- Its tail was long and unusually slim, resembling a bullwhip. A computer model detailed in a 1997 Discover magazine article found that the crack of the tail tip would have produced a "cannonlike boom" heard for miles. However, the slender tip of Apatosaurus' tail would probably be unable to hurt any predators, negating its use as a weapon.

Diet

- It's believed that Apatosaurus primarily fed on low-lying plants, but its long neck may have enabled this sauropod to eat soft leaves on higher trees, if its neck was flexible enough. Apatosaurus likely swallowed large chunks of plants without chewing and ingested stones to help with digestion.
- Like other large sauropods, Apatosaurus probably had to eat up to 880 pounds of food every day to survive, according to a 2008 study in the Proceeding of the Royal Society B.

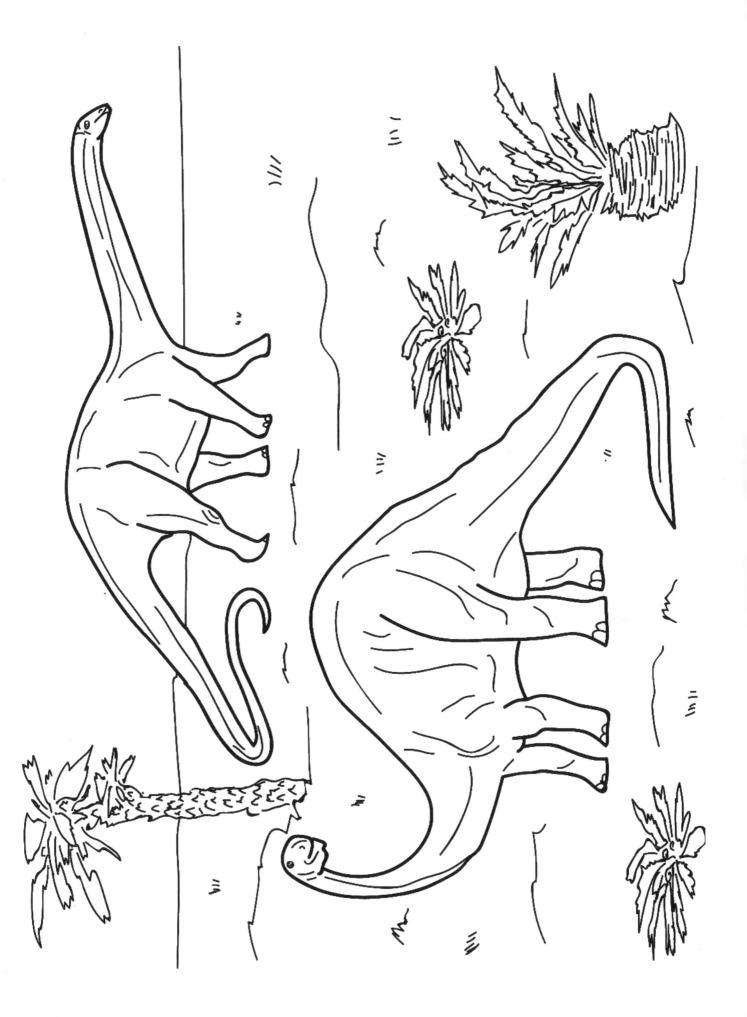

Elasmosaurus

Fun Facts

- One of the first identified marine reptiles, Elasmosaurus was a long-necked predator.
- The plesiosaurs lived in North America during the Late Cretaceous period.
- At close to 50 feet long, Elasmosaurus was one of the biggest plesiosaurs of the Mesozoic Era.
- Shortly after the end of the Civil War, a military doctor in western Kansas discovered a fossil of Elasmosaurus—which he quickly forwarded to the eminent American paleontologist Edward Drinker Cope, who named this plesiosaur in 1868.
- Plesiosaurs were distinguished by their long, narrow necks, small heads, and streamlined torsos. Elasmosaurus had the longest neck of any plesiosaur yet identified, about half the length of its entire body and supported by a whopping 71 vertebrae.
- Given the enormous size and weight of its neck, paleontologists have concluded that Elasmosaurus was incapable of holding anything more than its tiny head above the water.
- One thing people often forget about Elasmosaurus, and other marine reptiles, is that these creatures had to surface occasionally for air. They weren't equipped with gills, like fish and sharks, and couldn't live below water 24 hours a day.
- We don't know for sure, but given its huge lungs, it isn't unimaginable that a single gulp of air could fuel this marine reptile for 10 to 20 minutes.
- Judging by all those hoax photographs, you could make a case that the Loch Ness Monster looks a lot like Elasmosaurus.

Diet

- It was the long neck of Elasmosaurus that was key to its feeding method.
  All Elasmosaurus would have to do was swim up to a shoal of fish, possibly from below so that it could hide its body in the slightly darker depths, and use its neck to dart its head in and pluck out a mouthful of fish.

Deinonychus

Fun Facts

- It's not nearly as well-known as its Asian cousin, Velociraptor, which it played in Jurassic Park and Jurassic World, but Deinonychus is far more influential among paleontologists.
- The name Deinonychus (pronounced die-NON-ih-kuss) is Greek for "Terrible Claw" references the single, large, curving claws on each of this dinosaur's hind feet.
- Deinonychus could hold onto its prey with its fearsome front claws. Using its huge claw, it would kick and rip its prey apart. When not in use, the claw was held out of the way to keep it sharp.
- These animals lived in the Early Cretaceous, 120-110 million years ago.
- Deinonychus weighed around 165 lbs and were 10 ft long.
- In the late 1960's and early 1970's, the American paleontologist John H. Ostrom remarked on the similarity of Deinonychus to modern birds--and was the first paleontologist to raise the idea that birds evolved from dinosaurs.
- Paleontologists believe that most theropod dinosaurs (including raptors and tyrannosaurs) sported feathers at some stage in their life cycles. To date, no direct evidence has been presented for Deinonychus having feathers, but the proven existence of other feathered raptors (such as Velociraptor) implies that this larger North American raptor must have looked a bit like a large bird.
- Deinonychus had around 60 teeth and studies suggest that it had about the same bite force as an alligator.

Diet

- It was a carnivorous meat eating dinosaur.
- It isn't certain whether Deinonychus hunted larger dinosaur in packs - or jumped onto their backs and began feeding like some birds of prey do.

Archelon

Fun Facts

- Archelons were giant sea turtles known from fossilized remains found in North American rocks of the Late Cretaceous 100 million to 66 million years ago.
- They reached a length of about 12 feet long.
- Comparisons with the most similar and largest modern turtle, Archelon probably weighed over 3 tons (6,000 lbs).
- Archelon was a slow mover and found most of its food drifting near the sea surface. It had little need to dive deep except when hibernating on the seabed.
- Archelon's huge flippers suggest it was a long-distance swimmer in the open ocean. It would never be alone, as its huge size attracted a group of hangers-on such as juvenile fish as well as barnacles and parasites.
- Archelon couldn't withdraw its head or flippers inside its bony shell for protection so, despite its size, it was an easy target for large predators.
- Like modern turtles, it laid eggs by burying them in sandy beaches under the cover of darkness.
- Its nearest living relative is the world's largest turtle, the leatherback.

Diet

- Modern sea turtles are omnivorous, eating seaweed and algae as well as invertebrates and fish. Dermochelys lives predominantly on a diet of jellyfish.
- Archelon's great size probably meant that it could not chase down active swimming animals.

Ankylosaurus

Fun Facts

- Ankylosaurus was a heavily armored dinosaur with a large club-like protrusion at the end of its tail.
- Its tail was 10-feet-long and was likely used to swipe at the legs of predators.
- The huge tail could easily have broken the bones of most of its predators.
- Ankylosaurus means "fused lizard" in Greek, and it was given that name because bones in its skull and other parts of its body were fused, making the dinosaur extremely rugged.
- The top of the dinosaur was almost completely covered with thick armor consisting of massive knobs and oval plates of bone.
- Ankylosaurus lived in the late Cretaceous Period, about 65.5 million to 66.8 million years ago, and roamed the Western United States and Alberta, Canada.
- They would get to 35 feet long and 4 feet tall. Their width was about 6 feet and weigh up to 13,000 lbs.
- Ankylosaurs seemed to have had a large olfactory bulb (brain structure involved in the sense of smell), so the dinosaurs likely had a strong sense of smell to help seek out food and avoid predators.

Diet

- Ankylosaurus grazed on low-lying plants. The dinosaur's triangular skull was wider than it was long and had a narrow beak at the end to aid in stripping leaves from plants.
- Its small leaf-shaped teeth were not designed to break up large plants and it had no grinding teeth.

Titanoboa

Fun Facts

- Titanoboa was the largest snake ever, paleontologists have estimated that the body length of the average adult Titanoboa was roughly 42 feet and the average weight about 2,500lbs (1.25 tons).
- They lived during the Paleocene Epoch (66 million to 56 million years ago).
- This giant serpent looked something like a modern-day boa constrictor but behaved more like today's water-dwelling anaconda.
- Titanoboa probably spent much of its time in the water.
- It is extremely likely that Titanoboa had similar habits to the water dwelling anaconda because the animal's large size would have made living on land awkward or impossible.
- The presence of so many individuals displaying the same gigantic proportions demonstrates that 42-foot lengths were probably the norm for adults of this species.
- By comparison, adult anacondas today average about 21 feet in length.
- It was a swamp dweller and a fearsome predator, able to eat any animal that caught its eye. The thickest part of its body would be nearly as thick as a man's waist.
- It is estimated that Titanoboa may have had more than 250 vertebrae.

Diet

- Snakes are usually what you would term generalists that will eat whatever they can catch.
- Such possible prey items are crocodiles, lungfish, smaller snakes, turtles, birds, and mammals.
- Constrictors do not rely upon venom to subdue prey.
- By constricting, the method of killing is asphyxiation from the prey having its lungs squeezed by the muscular coils of a snake so that the lungs cannot expand to take in fresh oxygenated air.

Sarcosuchus

Fun Facts

- Sarcosuchus was by far the biggest crocodile that ever lived.
- The name Sarcosuchus is Greek for "flesh crocodile.
- Unlike modern crocodiles, which attain their full adult size in about 10 years, Sarcosuchus seems to have kept growing and growing at a steady rate throughout its whole lifetime.
- The largest SuperCrocs reached lengths of up to 40 feet from head to tail, compared to about 25 feet max for the biggest croc alive today, the saltwater crocodile.
- What made Sarcosuchus truly impressive was its dinosaur-worthy weight: more than 20,000 lbs (10 tons) for those 40-foot-long and 16,000 – 18,000 lbs (7-8 tons) for the average adult.
- If the SuperCroc had lived after the dinosaurs had gone extinct, rather than right alongside them during the middle Cretaceous period (about 100 million years ago), it would have counted as one of the largest land-dwelling animals on the face of the Earth.
- A full-grown SuperCroc would have been more than capable of breaking the neck of a large dinosaur, such as the fish-eating Spinosaurus, the largest meat-eating dinosaur that ever lived.
- Sarcosuchus was not just bigger than today's crocodiles, it was also a lot older. Most crocodiles have an average lifespan in the wild of around 25 years, with some individuals reaching 30 or more. Study on the growth rings present on some of the bones show that Sarcosuchus was around 40 years old and yet not fully grown when it died.

Diet

- Judging by the length and shape of its snout, though, it's likely that the SuperCroc ate fish pretty much exclusively, only feasting on dinosaurs when the opportunity was too good to pass up.

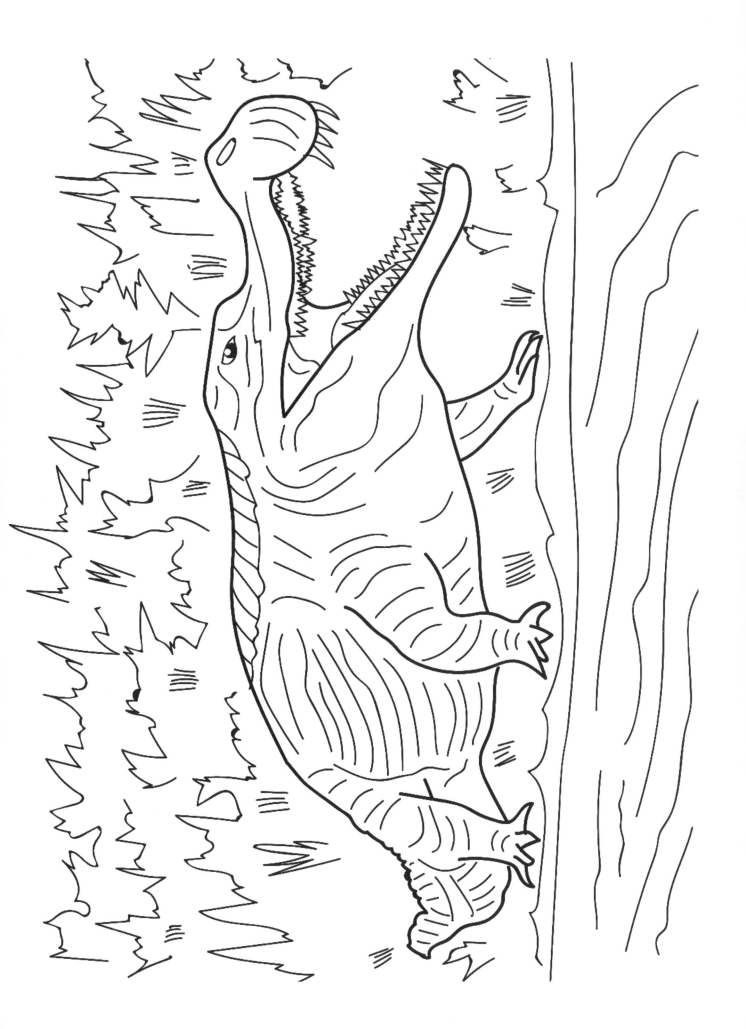

Megladon

Fun Facts

- The megalodon, which went extinct millions of years ago, was the largest shark ever documented and one of the largest fish ever on record.
- The scientific name, Carcharocles megalodon, means "giant tooth."
- Going solely by the size of the teeth, some believe that the fish could grow up to 60 feet long (18 meters), according to the Natural History Museum (NHM) in London, while others think that about 80 feet (25 m) long is more likely, according to Encyclopedia Britannica.
- Best estimates say their weight was around 60,000 - 100,000 lbs. (30-50 tons).
- It is important to remember that because sharks do not have bone skeletons, no Megalodon skeleton fossil has ever been found and these estimates are based almost entirely on their tooth size and by using the size of these teeth as a comparison to other sharks.
- Most of the fossils date back to the middle Miocene epoch to the Pliocene epoch (15.9 million to 2.6 million years ago). All signs of the creature's existence ended 2.6 million years ago in the current fossil record.
- Since there has been no evidence of the monster, including fossils that are any newer than 2.6 million years ago, most scientists believe that megalodons are extinct.
- The largest megalodon tooth measures around 7 inches (17.8 centimeters) in length, which is almost three times longer than those of great white sharks today.
- These sharks also had a ferocious bite. While humans have been measured to have a bite force of around 1,317 newtons, researchers have estimated that the megalodon had a bite of between 108,514 and 182,201 newtons, according to the NHM.

Diet

- The megalodon was a top-of-the-food-chain predator.
- It fed on big marine mammals, like whales and dolphins. It may have even eaten other sharks, according to Discovery.
- Researchers think the megalodon would first attack the flipper and tails of the mammals to prevent them from swimming away, then go in for the kill, according to the BBC.
- The megalodon's 276 serrated teeth were the perfect tool for ripping flesh.

The deadliest animals.
Average annual animal-caused fatalities in the U.S., 2001 to 2013

Sharks kill 1 person per year.

Alligators kill 1 person per year.

Bears kill 1 person per year.

Venomous snakes and lizards kill 6 people per year.

Spiders kill 7 people per year.

Non-venomous arthropods kill 9 people per year.

Cows kill 20 people per year.

Dogs kill 28 people per year.

Other mammals kill 52 people per year.

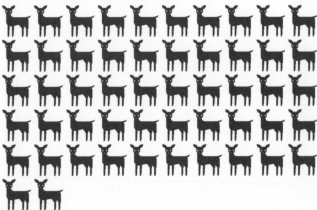

Bees, wasps and hornets kill 58 people per year.

WAPO.ST/**WONKBLOG**

Sources: CDC reports, CDC WONDER database, Wikipedia, Florida Museum of Natural History

As apex predators, sharks play an important role in the ecosystem by maintaining the species below them in the food chain and serving as an indicator for ocean health. They help remove the weak and the sick as well as keeping the balance with competitors helping to ensure species diversity.

What can you do to help wildlife?

- Reduce use of plastics
- Reusable drinking containers
- Using reef safe sunscreen
- Waiting 30 min after applying sunscreen to go into water
- Not standing or walking on coral reefs
- Keeping a respectful distance from wildlife
- Letting endangered sea turtles and monk seals rest peacefully on the beach
- Staying out of seabird burrowing areas because they are extremely fragile and could collapse with chicks inside
- Using wildlife friendly yellow and low frequency lighting outside to lower the risk of confusing seabirds and sea turtles that use the moon and stars to navigate
- Facing lights downward and using shields making them dark sky friendly
- Only have lights on when they are in use
- Make sure there is no standing water around your property to decrease mosquito breeding
- Do not feed stray cats
- Keeping domestic pet cats indoors
- Putting a bell on your cat helps to warn other wildlife but keeping your cat inside is better
- Keeping dogs on leashes especially near seabird beach breeding grounds
- Help honeybees by planting flowers that supply nectar and pollen throughout the season
- Plant native species
- Avoid using any poisons or chemicals outside
- Avoid using chemical fertilizers in your garden or yard
- Get out and vote for politicians that support and protect wildlife
- Join a conservation organization
- Consider a career in conservation/biology/environmental studies
- Seek out internships with environmental organizations
- Attend a beach cleanup
- Pick up trash every time you go to the beach
- Try to reduce carbon footprint
- Don't put hazardous substances down the drain or in trash
- Use cloth not paper napkins
- Recycle everything you can
- Don't leave water running
- Wash laundry using cold water instead of warm

Tyrannosaurus Rex

Triceratops

Utahraptor

Argentinosaurus

Kronosaurus

Stegosaurus

Pteranodon

Spinosaurus

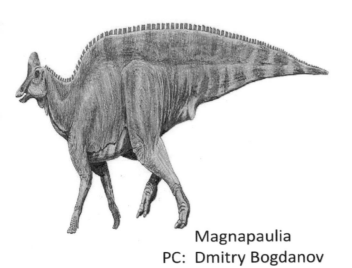

Magnapaulia
PC: Dmitry Bogdanov

Magnapaulia

Dunkleosteus

Apatosaurus

Elasmosaurus

Deinonychus

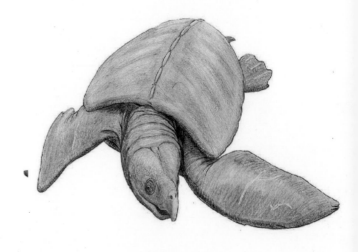

Archelon

Ankylosaurus

Titanoboa

Sarcosuchus

Megalodon